Chapter 1: <u>The Men Who Made Icons</u>

Remember when cars didn't all look the same? Remember when there was a time when companies had pride in themselves for being different? If the answer is no, then thats okay. You will enjoy this wonderful story all the same. There was a time when the focus was to make a rolling art that one could spot on the freeway from a mile away. The idea was to make a car an extension of the user. Like an arm to a body, the car would become a part of the person, his or her family, and then be passed down from father to son. Today, too many cars are designed without a soul, to fill a purpose of going from A to B. There were a handful of men who made this great era in automotive design possible. Two of them are named Richard Teague and Roy Lunn.

Richard Teague

Richard Teague was an industrial designer who worked for Chevrolet, Packard, Kaiser, and American Motors. He was born on December 23, 2923 and died May 5, 1991. In his early life he lost sight in his right eye in a car wreck. He held the position of Vice President of Design at American Motors Corporation from 1964 to the day he retired in 1983. Teague was known for making new designs on a short budget. Two of his designs were manufactured for two decades in one form or another. Lets visit the late 1960s. Richard and his design team were hard at work on a design that would become the AMC Hornet. The Hornet debuted in 1970. His clever design allowed the Hornet to be made in a 2 door sedan, 4 door sedan, a wagon, and a lift back coupe. Every 2 door sedan started life as a 2 door Hornet. His design allowed the factory to cut out and insert smaller, yet still accessible rear doors. Teague's ingenuity wasn't limited to different variations of the same car. The Hornet's design was also borrowed to produce the iconic AMC Gremlin. The Gremlin, which will be later covered, was essentially an AMC hornet from the B pillar forward. They sliced the car's design midway past the B pillar to quickly develop a subcompact car. In fact, they were very quick. Teague and his team were the first to create an American made subcompact car. American Motors beat the big three automakers to the punch by 6 months. The Ford Pinto and others were sure to follow, but the Gremlin, contrary to its given name, didn't have safety issues or mechanical issues that the other American car manufactures ran into by attempting to rush into the market. American Motors was so confident in the reliability of their little Gremlin that they offered the best warranty in the industry. By 1978, Teague and his team again altered the hood, front fenders, grille, headlights, and added more chrome to the Hornet. The result was called the Amc Concord which was a luxury compact car that was produced until 1983. In 1979, he and his design team sliced through the hornet design again. This time at a different angle. He also updated the interior with more plush materials. These two changes resulted in a new car called the AMC Spirit. The Spirt was also manufactured until 1983. Teague teamed up with Roy Lunn, a mechanical engineer, to make several variations of he AMC Spirit and AMC Concord. The sporty version of the AMC Spirit was called the AMX. That car ended up being the first American subcompact car to win at Germany's 24 hour legendary Nurburgring racetrack. Usually, the races were typically dominated by European manufactures. Teague and Roy Lunn also designed a 4wd version of the Spirit and Concord. They dubbed it the AMC Eagle. The eagle was produced until 1988. Richard Teague and his team's creative ability to stretch the lifespan of a design makes him arguably one of the best American industrial designers to live. His late 1960s compact car design ended up being manufactured in one form or another from 1970 to 1988. Richard Teague was also named "Man of the Year" by

"Chilton's Automotive Magazine" in 1976 for his work on the AMC Matador design. Some of Richard Teague's iconic designs are the following:

AMC Javelin

AMC AMX

AMC PACER

AMC Gremlin

Jeep Cherokee

HMMWV

Roy Lunn

Roy Lunn was an aeronautical and mechanical engineer who worked for Aston Martin, Ford Motor Company, AM General, AMC, and Jeep. Roy was born in 1925. He obtained the position of Vice President of Engineering of AM General. Roy Lunn is responsible for engineering several icons. His first icon is the Aston Martin DB2. Aston Martin Fanatics commonly pay over $500,000 for one of these rare beauties. He engineered the Ford Mustang I concept, the Ford 429 Boss Mustang, the original Ford GT40, and many other cars.

When he began working for AMC, he spent a good amount of time working for the Jeep devision. Some of his notable engineering achievements at AMC include the Jeep Cherokee, the Jeep Comanche, the 1980 Spirit AMX , and the Eagle.

He designed his most renowned work after he retired. American Motors called back both Richard Teague and Roy Lunn to design the High Mobility Multipurpose Wheeled Vehicle also known as the military hummer. The HMMWV is still being used by the military today. The design was so popular that AM General, one of the surviving branches of AMC, sold licensing to General Motors to make a civilian model. General Motors dubbed the civilian model the H1. The success of the H1 led to the development of an H2 and then an H3 as well. A pickup version of the H3 was even produced!

The rest of this book is dedicated to the products that either Richard Teague, Roy Lunn or both together had a hand in creating.

Chapter 2: Roy Lunn's Solo Career

Roy Lunn worked on many projects. Thats good thing. We were lucky he went into the automotive industry. Noted here in this book are the Aston Martin DB2, Ford Mustang I Concept, Boss Mustang 429, and Ford GT40.

Aston Martin
DB2

1950-1953
- Duel overhead cam 2.6L Straight six 125hp (93 kW)

- Finished 3rd in the Spa 24 hour race

- Claimed the life of race car driver Pierre Marechel

- 411 were produced

- 1947 David Brown Buys Co.

- First 49 had a three piece Chrome0framed front grille

- Top speed 116 miles per hour

- 0-60mph in 11.2 seconds

- Average price today is $500,000

Ford Mustang I

1962

- Mid Engine Design

- 60° V Style 1.5 L 4 cylinder Engine (109 hp)

- Front Wheel Drive Powertrain

- 155 bhp per tonne (1-60 in 11 seconds)

- Only one concept car was manufactured

- Design inspired car designs like the lotus Europa and Fiat X/19

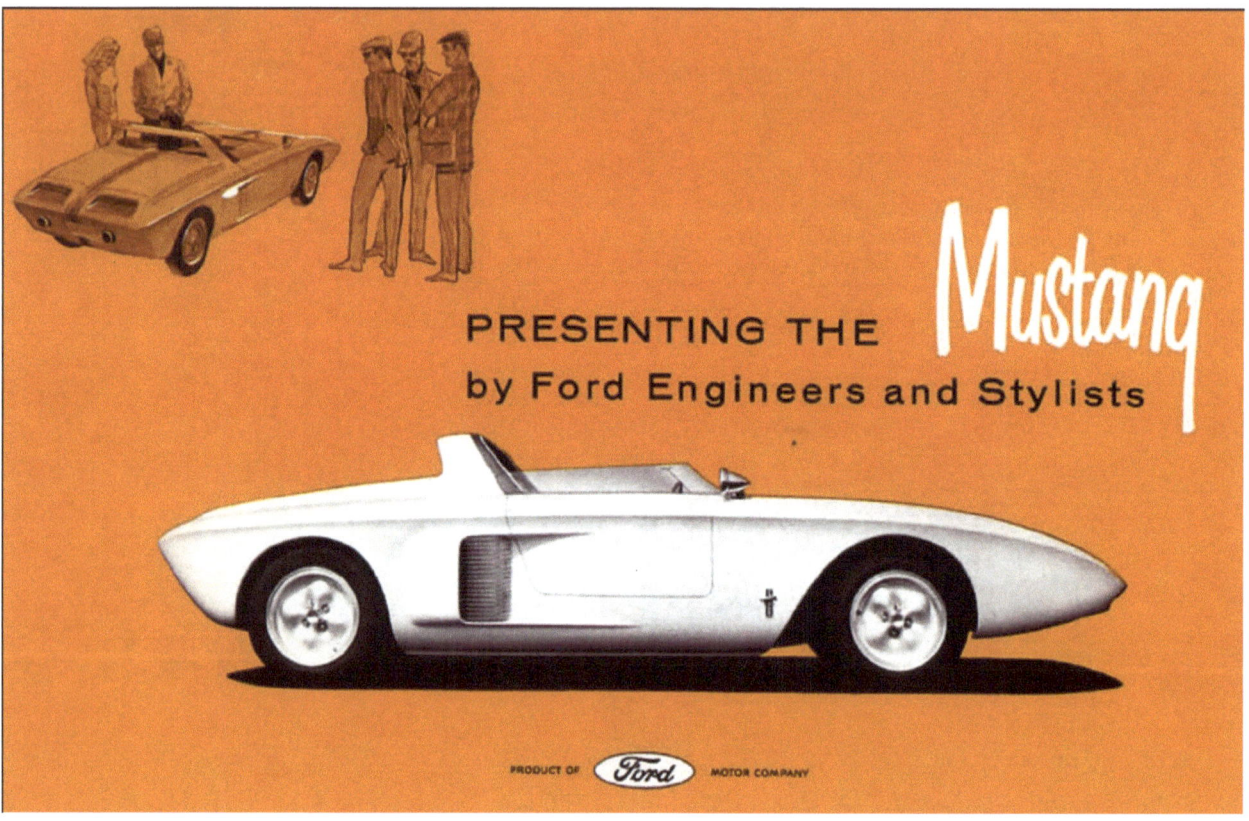

- Prototype designed and finished in 100 day

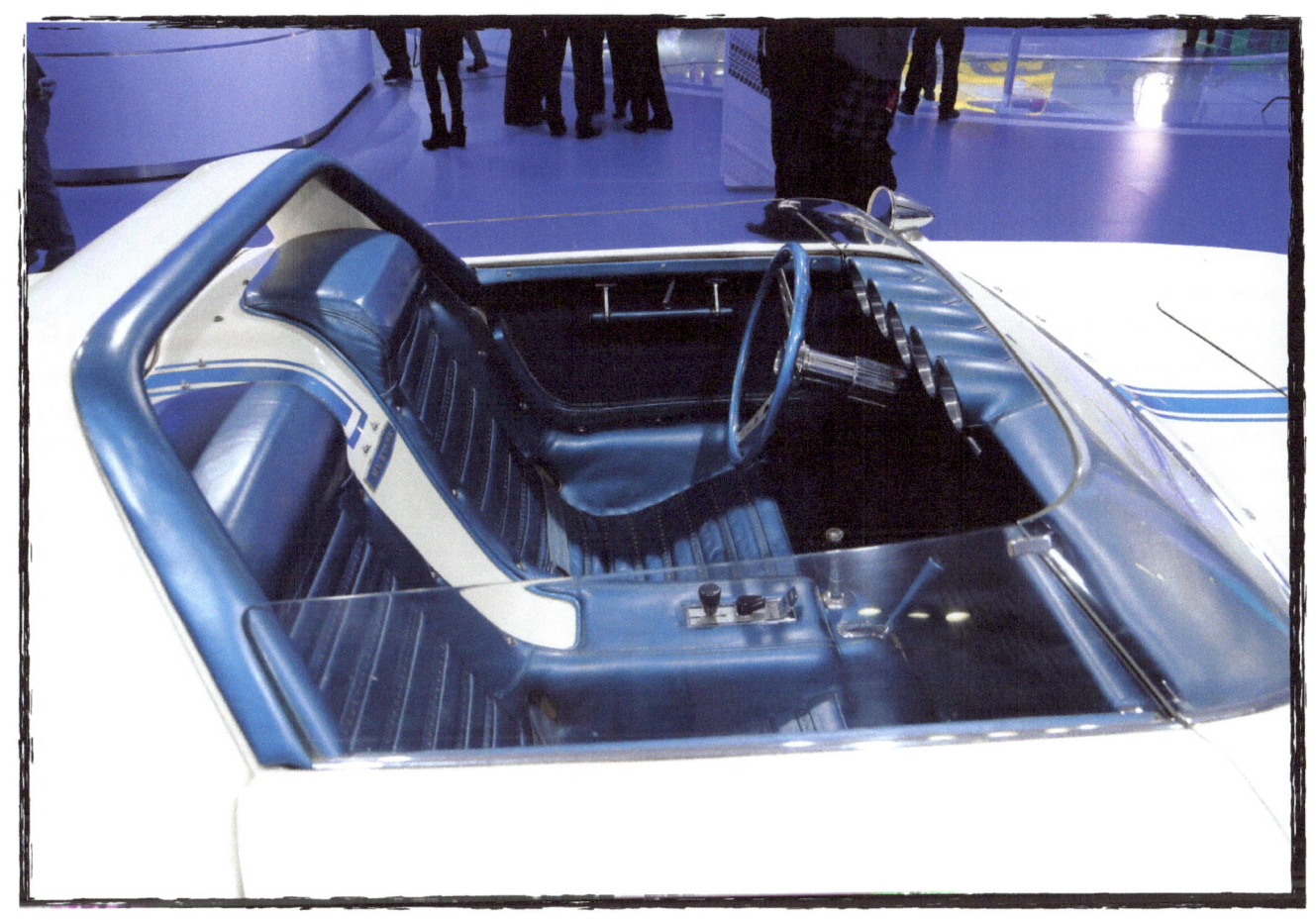

- Roll bar was integrated into the body panels

- Seats were embedded in the body. However, the pedals were adjustable

- Ford I concept badge is very close to the actual production mustang badge

- The car featured a 4 speed transaxle

Ford 429 Boss Mustang

1969-1970

- 859 in total were produced
- Rear Wheel Drive
- Cougar variant was produced as well
- Designed to compete with dodge in the racing circuit

Hemispherical designed engine

Engine was banned from NASCAR

375hp and 450 lb ft of torque

8000 rpm tachometer

FORD
GT40

1964-1969

Ω 100 Produced

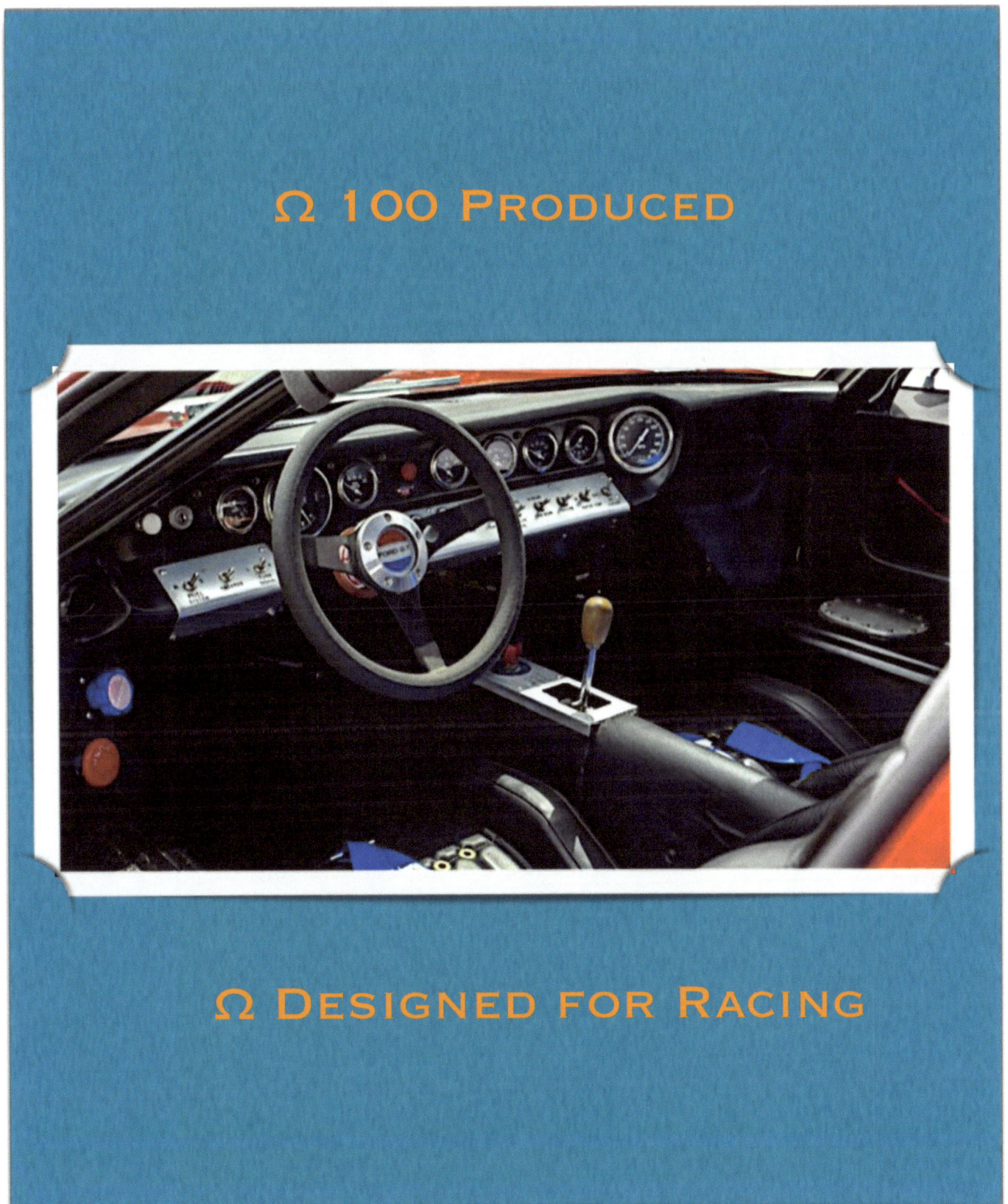

Ω Designed for Racing

Ω Won 24 Hour Le Mans Race 4 Consecutive Times

Ω Won 1966 and 1967 International manufactures Championship

Ω Fabricated in England

Ω Mid Engine Design

Ω 4 Speed transmission

Chapter 3: RICHARD TEAGUE'S CONTRIBUTIONS

American Motors
Javelin AMX

1968-1970

Motortrend: "Most notable new entry in its class" and top "sports-person category" and the top choice for "sports-person category"

A concealed roll-bar was built into the cantilever of the roof

First manufacture to use safety padding. This was a full plastic dash that increased crash protection.

Craig Breedlove set 106 world speed records

Second place in the 1970 trans Am Championship and the second generation won first place in 1971, 1972, and 1976

The car had a top speed of 140 mph and went from 0 to 60 in 6.9 seconds.

The car was powered by a Rambler designed 390 v8 engine which was not the ford 390 v8 that many confuse the engine with.

American Motors
MATADOR

1974-1978

The matador was an aerodynamic design an area of boxy cars

Design made Richard Teague Man of the Year in 1974.

Car and Driver compared the work to the artist Charles Eames.

Car was meant to compete in the personal luxury car segment

The car was featured in the James Bond Film "Man with the Golden Gun"

Car featured several engine choices ranging from a 258 I6, a 304v8, a 360v8, and a 401 v8.

The car faired well in the NASCAR racing circuit. However, the car failed to sell well.

AMERICAN MOTORS GREMLIN

1970-1978

There was a race between the domestic automakers to produce cars to compete in the subcompact car market. American motors won, with th design of the gremlin. Design process for the car was shortened dramatically when Teague took an AMC Hornet and sliced the design at the B piller.

This slicing technique allowed for many shared parts and resulted in a differentiated product that really was identical to the hornet from the B-pillar forward. Although platform sharing is used by almost all major manufactures today, back in 1969 this was a new concept. Amc, and Richard Teague was truly ahead of its time.

Teague was criticized harshly because the gremlin was first manufactured with a small rear window that was hard to see out of. He responded and redesigned the window that appeared on the 1977 year model

Like the 2 door AMC hornet, the 4 door hornet, the 2 door coupe, and the hornet wagon, the Gremlin was offered with a 232 I6, 258 I6, and a 304 V8. The car was also offered with an I4 Audi engine

The gremlin sold well for a car that was initially sketched on an airplane sickness bag. "Newsweek" magazine featured the ncw gremlin on the cover with the caption "Detroit Fights Back." even though the car was manufactured in Wisconsin.

Bob Nixon deserves some credit as well, because he also helped finish the design of the gremlin.

AMC
PACER

1975-1980

- Compact car that was as Wide as a full size car of the same era.

- First Cab Forward Design which is now used on almost all modern vehicles

- Designed to be roomy, aerodynamic, and fuel efficient. Compact cars today now follow the same set of standards that the Pacer established

- The car, due to its "ugly styling" unfairly continues to be on worst cars ever made lists.

- Sold as the car of the future that you could own today

- Sold with 6 and 8 cylinder options

- Built to exceed safety standards of the day

- Car was featured in the movie Wayne's World

- Despite claims of poor reliability, Consumer Reports concluded that the car scored higher in reliability than any other domestic subcompacts of the era.

American Motors AMX/3

1970 Experimental Design was first penned on a cocktail napkin

- 6.4L - 430 lb*ft -340 hp

- Mid Engined Design with a 4 speed transaxle

- Hand Built by Giotto Bizarrini

- $2,000,000 spent in development

-     BMW stated chassis was one of the stiffest and had the most neutral handling they have ever tested

- 6 cars were produced

- Top speed 170mph

- last one on the market sold to a collector for $795,000

- Debuted one day before the Ford Pantera

Chapter 4:
RICHARD TEAGUE'S AND ROY LUNN'S JOINT PROJECTS

AM General

HMMWV

1985-Now

- 15 Configurations

- 16 inches of ground clearance

- 6 feet tall by 7 feet wide

- 24 Volt Electric System

- 3-speed Transmission

- 7 mpg

- Chosen from 11 other prototypes

- Designed to be a light transport vehicle

- Civilian version was manufactured by

AMERICAN MOTORS
1979-1980 Spirt AMX

- First U.S. automobile to win in its class at Germany's 24 hour legendary Nurburgring racetrack. The movie Ultimate Challenge was made to tell the story of this event

- There was an emphasis on corrosion engineering that lead to galvanizing the steel bodies, creating the bumper out of aluminum, and making the spoiler out of fiberglass. Lap joints were minimized, and rust protective coatings were applied to the undercarriage of the vehicle.

- The AMX Shared Body Panels with the AMC Spirit which was a facelift gremlin with a redesigned rear.

- There were 5000 produced. The car could be ordered with a 4.2L I6 engine or a 5.0L v8. The engine was considered to be the lightest inline 6 engine in the industry. The car could be mated to a three speed auto or a 4 speed manual transmission.

- The AMX was designed to compete with the Ford Mustang. The car did not sell well despite its racing success

Jeep Cherokee XJ & Comanche MJ

1984-2001 Cherokee

1984-1992 Comanche

The Jeep Cherokee and Comanche was Richard Teague's swan song. At the time, Roy Lunn was the head of Jeep engineering, Richard Teague was close to retirement, and American Motors had merged with Renault. The thought was that AMC and Renault would not only share technology, but they would sell Jeeps across their newly acquired dealerships throughout Europe and Renaults across America. One early concept of a 2 door Cherokee had some French styling.

The Jeep Cherokee and Comanche used Renault designed throttle body fuel injection system for the 2.5L I4 engine, and a Peugeot 5 speed automatic. A 4 speed manual was offered for the base model and a 4 speed auto was offered as well.

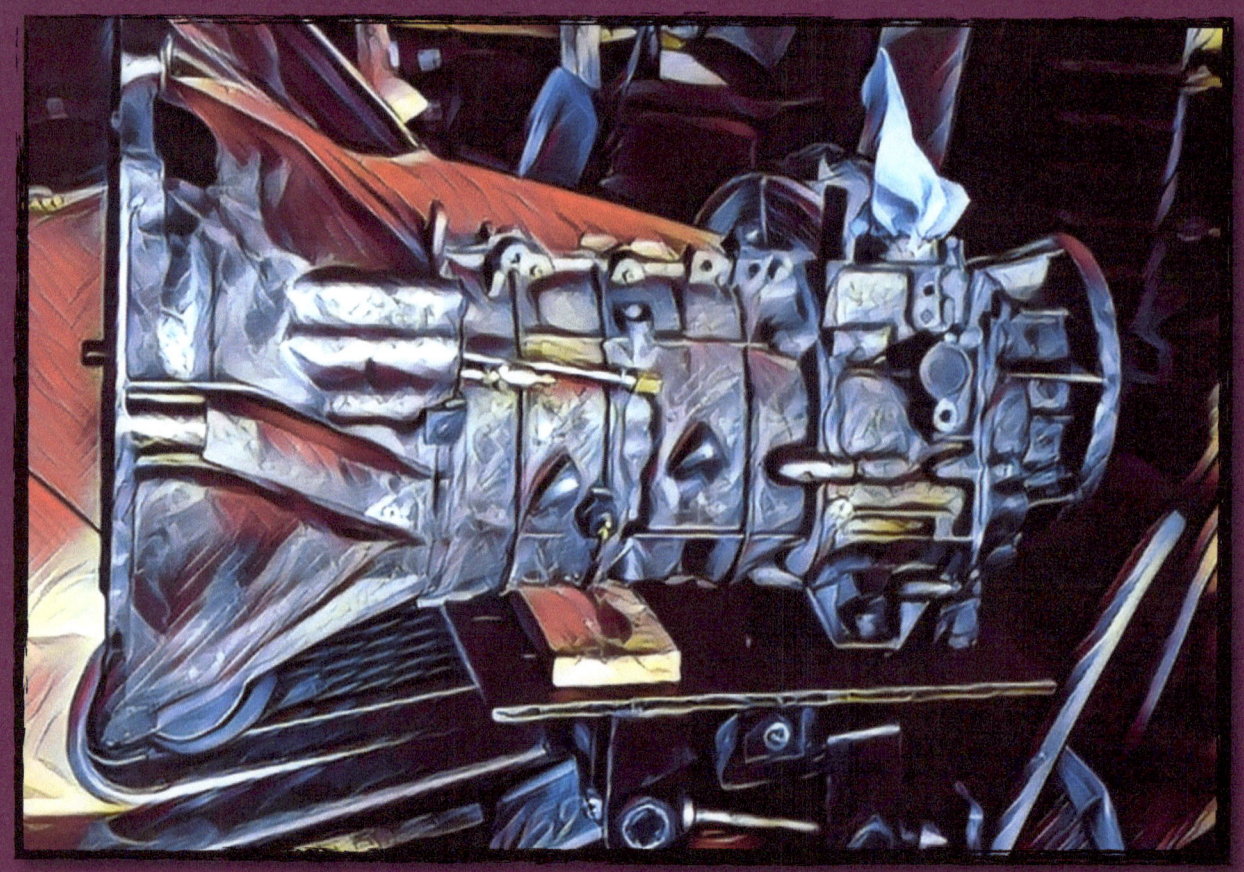

A Chevrolet designed v6 was initially offered as an option as well. However, the v6 didn't offer much of a power advantage over the I4, and the rear main seals on the v6 leaked on the showroom floor.

The steering column, key, and lock and tumbler were also supplied by general motors. Meanwhile, the distributer was designed by Ford Motor Company. Even though the Jeep Cherokee could be considered an agglomeration of "just everybody else's parts," The little 4x4 had some great attributes

The Cherokee was Jeep's entry into the compact SUV market. Well, actually, it's considered the first SUV on the market. The Cherokee offered 4 doors in an era of 2 door off roaders, modern styling, modern engineering, a quiet cabin, a smooth non industrial equipment sounding transmission, and shift on the fly 4 wheel drive. This shift on the fly feature allowed the user to shift into 4 wheel drive without stopping the vehicle or turning a set of hubs in. Instead, the hubs were locked via vacuum when the 4 wheel drive lever was used.

The Cherokee was a resounding success. The vehicle on all three major automobile magizines' four wheeler of the year award. Over 2 million Cherokee's were manufactured between 1984-2001. The suspension design was so good that when Chrysler absorbed Jeep in 1987, they used scaled up the suspension system to work on the mid 1990's ram truck and even the 1990's grand Cherokee. Jeep even further refined the engine choices. American Motors replaced the General Motors v6 with a port fuel injected 4.0L Inline 6 that had similar power to AMC's own 360 v8 at the time.

American Motors hired metallurgists, (material scientists) and tribologists (wear engineers) to make the new inline 6. They had a limited budget so the old 258 I6 designed in the 1960s was used as a base. This wasn't a bad thing however. It was the lightest inline 6 on the market. It also gave the design engineers plenty of wear data to work with. The over 20 year production run of the old engine provided engineers with the opportunity to know what worked well, and what wore out over time.

The block ended up being case hardened via a carburization process. Tolerances were decreased, surface finishes were finer, and engine noise was decreased. The head, intake manifold, and exhaust manifolds were redesigned to be lighter, and increase airflow.

American Motors needed a win. They needed to get this car right. They knew that if AMC was going to survive, it needed a knockout puncher. Sadly, all of this effort didn't save the company. It did get the attention of Lee Iacocca. He directed Chrysler to buy American Motors for the Jeep brand.

The Comanche, did not sell as well as the Cherokee, and was axed because it competed directly with Dodge's own Dakota pickup. Kiplinger placed the platform on the "10 cars that refuse to die" list, the 4.0L is still considered the best Jeep engine of all time, and the Cherokee is still a popular choice for many off roaders.

The cars featured here improved peoples' lives, changed the culture of the American people, and made their way into the hearts of many enthusiasts that restore these cars. Thank you for reading this book. It was a blast to make, and feel free to contact me at zpjstewart@gmail.com if you think you have a better photo of one of these cars.

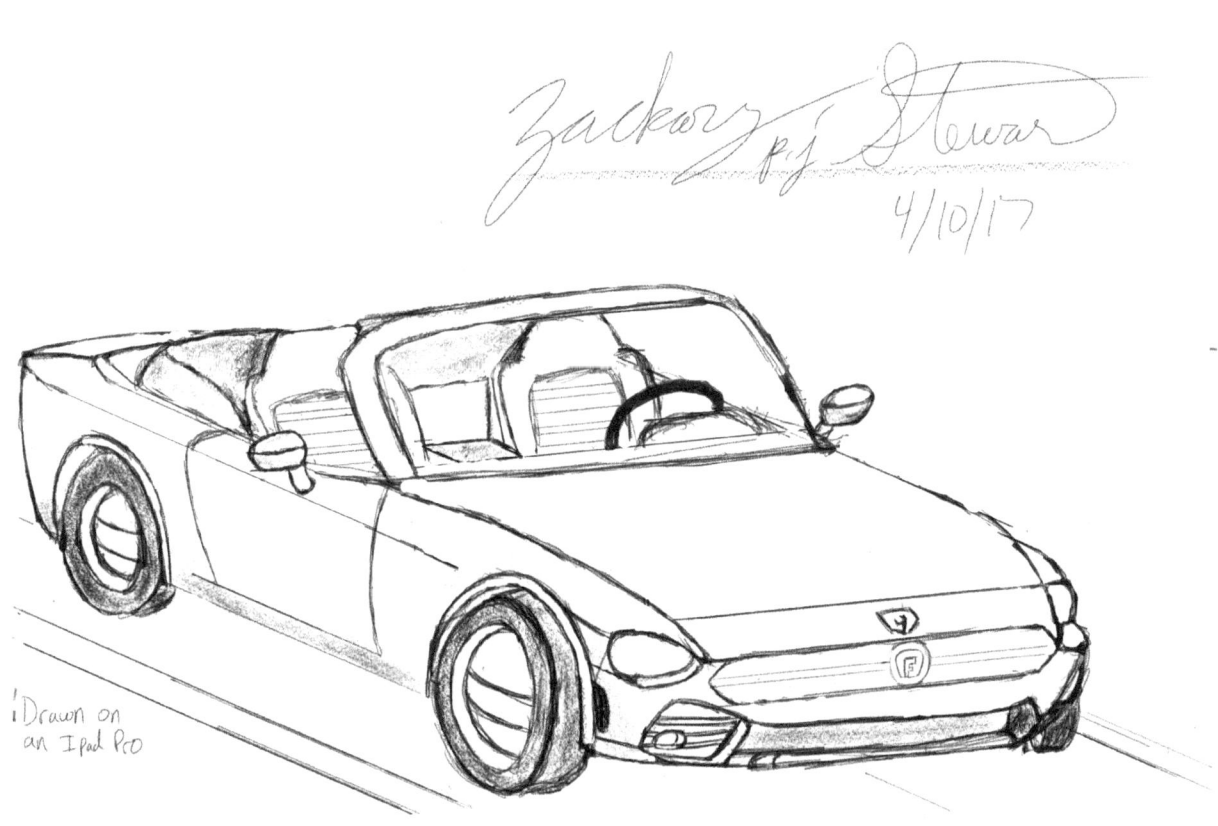

With God all things are possible

Matthew 19:26

References

DB2 By: Nick Aaldering

1970 AMX photos by: Vanguard Motorsales.com Muscle Cars and more

1980 AMX photos by: Zackary PJ Stewart

AMC Gremlin photos by: Vanguard motorsales.com Muscle Cars and more

AMC Pacer Illustration by: Zackary PJ Stewart
AMC Pacer photos by: Vanguard motorsales.com Muscle Cars and more

AMC Matador Photos By: Kevin G

Jeep Comanche Photos by: Zackary PJ Stewart

Jeep Cherokee photos by: Zackary PJ Stewart

AM General HMMWV photos by: Zackary PJ Stewart

AMX/3 illustration by: Zackary PJ Stewart and Jason Clark

Ford Concept photos: By Bzuk at English Wikipedia - Transferred from en.wikipedia to Commons by Liftarn using CommonsHelper., Public Domain, https://commons.wikimedia.org/w/index.php?curid=4488257

www.ingramcontent.com/pod-product-compliance
Lightning Source LLC
Chambersburg PA
CBHW041315180526
45172CB00004B/1104